The Black Hole

By Forester de Santos

Something about the Black Hole

The Black Hole is a giant hole or a round rupture or tear or breakage in the fabric or the foreskin skin which really is the vacuum of space.

But the black hole or the tear or the breakage in the vacuum of space moves along the vacuum of space sucking up or removing or destroying all matter, including all the stars and planets and even all forms of life, to clear the universe or the vacuum of space for new matter or for another beginning which will appear as if the very first beginning because there will not be any trace that there was ever a first beginning...

Tags: black hole, vacuum of space, stars, planets, beginning, universe, matter, breakage, end, foreskin, cosmos...

About this Author Beloved

All writers must come sooner or later to the things that they want to truly write about and most come to write where the easy money is and most writers make a very big killing in becoming very rich and very famous by writing fiction and good for them!

But the great question is how much fantasy for the human race since the human race has being living in a fantasy since the race entered into consciousness or began to think and invent with words?

Well, that is a great or tall question that Forester de Santos as a writer has truly asked!

And so he has truly chosen to walk or to actually write on the road less taken or less written about!

And so he began to research and write about immortality and as the same goes, one becomes what one thinks or writes or even reads about the most!

Prologue

Existence truly is about knowledge not only to be born alive and to survive but also existence truly is about acknowledgement to be reborn alive in life, which really is entering into a higher or a taller mental consciousness or receiving an expansion of mental consciousness.

Acknowledgement

I acknowledge my existence because he is a tall genius and he has also made me a genius because of me presenting him for a tall genius...

Dedication

The Black Hole is dedicated to all of those that really seek the truth so that the truth frees them and they thus truly become for much more for they truly becoming for the very truth herself through a higher or taller mental consciousness…

Table of Contents

Introduction

The Black Hole is a giant hole or a round rupture or tear or breakage in the fabric or the foreskin which really is the vacuum of empty space.

But the hole or the tear or the breakage in the vacuum of space moves along the vacuum of space sucking up or removing or destroying all matter, including all the stars and all planets and even all forms of life, to clear the universe or the vacuum of space for new matter or for another beginning which will appear as if the very first beginning because there will not be any trace that there was ever a first beginning...

But no need to worry because the above matter will happen billions of years in the future. By the time that our galaxy is intersected by a black hole for destruction, our sun has extinguished for lack of fuel or energy and life on our planet has also stopped from existing because of lack of illumination...

Thus, the true purpose of the black hole, however, is to turn the vacuum of space or the universe into a neutral point (0) to attract new matter or new energy from outside of the vacuum of space or from outside the universe which will become the new stars and the new planets of the new universe after the new big bang!

But as long as there is a single black hole in the vacuum of space or in the universe then the vacuum of space or the universe will be negative or negatively charged (-) and nothing will enter the vacuum of space or the universe as long as the vacuum of space or the universe remains negative or negatively charged...

Chapter One

The Black Hole

The heaviest or the denser matter or the heaviest or the denser stars at one point in time or at one point in the vacuum of space will turn into supernovas and then into super black holes or turn into the cancer cells of the universe and universes, yes there are other universes or other dimensions of times!

Actually, black holes or collapsed stars are like giant vacuum cleaners or giant space sweepers, really giant magnetic graves!

There will be another point in time or another point in the vacuum of space when most of the universe and universes will be occupied by super or by giant black holes or by giant cancerous stars or dark stars or vast empty magnetic graves.

Once again, the universe or universes will be occupied by giant vacuum cleaners or giant sweepers or giant graves.

The universe and universes will look like a giant Swiss cheese, but without the appetizing smell.

At that very stage, when the entire universe and all universes are occupied by enormous black holes or enormous dark stars or enormous vacuum clears, the enormous black holes or the enormous dark stars will disintegrate or will drain the entire light or the entire energy or the entire matter of the universe and universes spontaneously or at once!

That is to say, the giant black holes or the giant space vacuum clears will begin contracting or moving or going to the center of the vacuum of space or to the actual physical point of impact or to point zero or to the point of origin until the entire physical universe and the other universes collapse—the exact opposite of the Big Bang! Or the Big Bangs, all 118 of them!

The black holes or the collapsed stars will begin to move toward the center or begin moving toward the actual physical beginning or the physical origin of the universe and universes, removing all the matter, all the energy or all the light from the entire universe and universes and thus making one giant or making one enormous black hole or one giant or one enormous vacuum per universe.

It is this new giant or this enormous black hole vacuum in the vacuum of space or inside the magnetic sphere which is the universe that actually attracts new matter, new energy or new light from outside the universe or the vacuum of space.

Once new matter, new energy or new light from the outside of the universe or outside the vacuum of space enters or re-enters the giant black hole vacuum in the vacuum of space or in the in the universe, the new matter, the new energy or the new light is super compressed recreating the Big Bang! Or really creating the Big Bangs! And thus destroying the giant or the enormous black hole vacuum!

Actually, every giant or every enormous black hole in the vacuum of space or in every occupied universe or in every occupied dimension of time or in every occupied dimension of space will be destroyed at once or at the very same time by new matter from outside of the universe or from outside the vacuum of space!

In other words, every giant black hole vacuum occupying the 118 universes or occupying the 118 dimensions of time will be destroyed at the very same time by new matter, new energy or new light coming from outside the universe or outside the vacuum of space.

However, there also exists the possibility that the giant black hole or the black holes will disperse as if a storm and space will become as if a sea of tranquility or neutral (0).

Chapter Two

The Stars and the Black Holes

The universe or the vacuum of space is a neutral point or is a point zero which is converted into a positive point or into a one or more when light or matter or energy enters into the universe or into the vacuum of space from outside the universe.

The universe or the vacuum of space also is converted into a negative point when the grand majority of light turns off or the majority of matter no longer has energy or fuel...

Before light or matter enters into the universe or the vacuum of space the weight of the universe is cero and when light or matter enters into the universe thus the universe takes on weight.

Light or matter is composed of 118 Elements and the weight of each element is two times its atomic number, more like its positive number.

In the case of element number one, for example, its weight is of two and in the case of element number two its weight is of four and in the case of element number 118 its weight is of 236...

Curiously, that the totality of the 118 elements adds to one and that the totality of their weight adds to 2.

That is to say, that if we added from one to 118 thus the sum would be 7,021. And if we added that sum thus it would be 10 and if we add that last sum thus it would be one. That is, 118 is equal to one...

And if we added from two to 236 the sum would be 14, 042. And if we added that sum thus it would be 11 and if we added that last sum thus it would be 2. That is, 236 are equal to two...

Thus, light or matter enters into the vacuum of space or into the universe as one or as a unit which is composed of 118 pieces or elements and the weight of the element is two times the atomic number of the element.

Thus in truth, number one itself is composed of 100 percent plus 18!

That is to say, that the number one or even oneself is equal to 118 percent!

Also light or matter could enter into the vacuum of space or into the universe with only or as one element with its weight of two but that element would be able to convert into the other elements, even to the element 118 and its double weight of 236...

Element number one, for example, enters into the vacuum of space with its weight of two and it has 117 other possibilities of converting into the other 117 elements.

That is to say, element one is composed of 118 parts or pieces or the 118 percent and element number has 117 possibilities of converting into the other 117 elements according to the weight which element number one maintains.

In the same manner as above, element number two with its weight of four has 116 possibilities of converting into the other 116 elements or until the element number 118 with its weight of 236...

When light or an element enters into the vacuum of space, light or the element enters as if it were a piece of magnet or as if it were a bar magnet, but in the form of a cube.

In the vacuum of space the magnet or cube or light or the element is super compressed not only until it takes the form of a sphere or round but also light or the element or the magnet is super compressed until it gets to a very high level of temperature.

And when the temperature gets to its highest level thus light or the magnet or the element super explodes causing the light or the magnet or the element to divide into two parts and the part with less weight, such as the negative part, takes position in the magnetic field and that magnetic field now is a negative field or the electron ring...

The superior or the positive or the heaviest part of light or of the magnet or of the element takes position in the center or in the nucleus.

Thus, now we have the negative part of the magnet going around the positive part when before they were united and the center or the nucleus was neutral...

And even though the element lost weight because of the explosion or because of bursting, the element continues the one for two.

That is, its weight continues of two although the element one now is 0.999 and its weight is double, 1.998…

Thus, now element number one was reduced to about 0.999 with its new weight of 1.998 but to element number one also remains a neutron or even more than one or a neutral part which can be converted or can be transformed into a positive part and that way not only adding the number of the element but also adding its weight even though it will have only one negative particle going around the center or the nucleus or the positive side…

But if one or light or the magnet or the element does not convert into the next number or into the next element thus it loses its energy and will only be a piece of dead matter in the vacuum of space and it will be removed one day by the black holes…

Thus, light or the element is the same as a negative particle, is the same as a neutral particle and is the same as a positive particle.

In a way, one is equal to a negative portion plus a neutral portion plus a positive portion which totality is of 0.999 after entering the vacuum of space.

But in the vacuum of space light or the element is positive even though the vacuum of space is neutral but reacts as if negative because of the vacuum or compression.

And when the element increases its positive part by converting the neutral part into a positive part, the element cannot attract the negative part because of the vacuum of space because now the negative part becomes as if more or its weight increases because of the weight that it receives indirectly from the vacuum of space.

And if there were not positive attracting the negative thus the negative would expand through the vacuum of space and it would stop being negative and it would be dead matter...

Now, an element, in this case a star, which number is high as the same as its weight thus lasts or remains longer in the vacuum of space or in the universe.

But the element or the star becomes heavier while the energy or matter to continue on lasts or it begins to transform from positive to neutral and once neutral, the element or the star practically becomes negative when its excess or super weight attracts the electrons toward the center and thus causing an implosion in where the element or the star becomes a nova or a new star but without light or without energy and that way causing an enormous hole in the vacuum of space when before the element or the star occupied the vacuum of space as an element or as a star...

In other words, matter or the star in the vacuum of space changes from positive to neutral and then from neutral to negative.

Thus, + 0 -, in where the negative is a black hole or a super vacuum cleaner in the vacuum of space.

This black hole practically eats or sucks all the matter around it to take all matter out from the vacuum of space or from the universe to make new space for new matter or for another beginning...

But as long as there are black holes in the vacuum of space or in the universe, thus the vacuum of space or the universe is negative and as long as the vacuum of space or the universe continues as or is negative thus it keeps being for something and not for neutral and from neutral or from zero to positive...

Thus, so that the universe becomes neutral or to zero and from neutral or from zero to positive thus all the black holes or super space vacuums must stop from functioning and once the black holes or the super space vacuums stop from functioning for lack of matter or for lack of energy thus the vacuum of space or the universe will return to neutral or to zero...

Now then, this new vacuum in space or in the universe is neutral and has zero energy but even so attracts new matter or new stars or attracts the new or the next beginning which is outside of the universe or outside the vacuum of space...

And once new matter or new stars enter into the vacuum of space or into the universe thus the vacuum or the universe will be positive or one or more...

Thus, the cycle or the model or the rhythm of (+ 0 -) continues until the end of all the times...

Chapter Three

Something More About Existence

Existence exists because she cannot ever stop from existing. Existence has no choice but to only exist and exists for all eternity.

Existence really is composed of three portions of lack but which can be seen and existence also is composed of one portion which is but cannot be seen as of yet and which makes for three or even makes for four or for more portions...

Existence also is composed of opposites which really are more opposites of themselves than to the opposites to which they are opposites of.

And this opposition or contrast which makes possible for existence to attract herself and it appears that existence moves or is in motion because of the very attraction of herself.

And existence is known or is seen more through the little which is seen of her and through the more which is seen of her, less of her is known or less is seen...

And even though existence exists and she attracts herself and she also is eternal knowledge and also she gives knowledge to the vacuum of space or to the universe and thus creating the beginning or the times, existence does not know that she exists even though she also renews herself so that she can continue existing and existing as if always new and as if nothing has happened and will never happen...

But existence renews herself or is reborn when the vacuum of space becomes empty again or when the knowledge given to the universe is not taken as acknowledgement and she once again gives or once again throws more knowledge into the vacuum of space or into the space of the universe, thus making as if a new beginning and everything which was before as if it never had being or as if it never were...

But that ability which existence has of renewing herself and of continuing on for all eternity without knowing end or ending thus we conscious beings also possess.

But we are not obligated to live eternity as neither we are obligated to death or to the end or to the ending.

Death or the end or the ending comes because it is not done to continue on but to continue on as if new and to continue on forever or to continue on for all the times...

Those that do not want eternity thus they just wait for death and death will do for them so that there is no eternity for them...

But those of us that want to continue on with life without seeing or without knowing death thus we must do to be reborn and that rebirth or being born again truly is done with knowledge or with acknowledgement from above.

Thus, we must give knowledge or give acknowledgement to that grandiose part of existence which is but which cannot be seen as of yet but even so it is what gives or it is what grants knowledge not only to the vacuum of space or to the universe but also it is what gives or what grants knowledge to every conscious being that requests it...

And just as one presents oneself to that grandiose part of existence which gives or which grants or which even lends knowledge, thus that grandiose part of existence will be to one.

In other words, that grandiose or glorious part is the Creating part of existence or the renovating part or the part which gives identity to all existence or which really reacts with all the opposites of existence...

Thus, he that gives or that grants or that lends to the Creating part or to the Renovating part of existence the knowledge or the acknowledgement of a higher consciousness or of God or of Creator or of Renovator thus he also will have the knowledge or the acknowledgement of a higher or taller consciousness or of God or of Renovator and God will present to him or come to him or will allow him to draw near with that very same knowledge or acknowledgement...

Existence gives or grants or lends to the vacuum of space or to the universe knowledge in the form of matter or stars or light or the element.

And if that knowledge or matter in the vacuum of space or in the universe really is converted into acknowledgement thus the universe will continue on as if forever new without ever knowing end or ending nor knowing or remembering beginning...

But if that knowledge given or granted or lend to the vacuum of space or to the universe does not renew or is not converted or is not transformed into acknowledgement, thus that knowledge in the form of matter or stars or light thus will lose her energy and there will only remain and will be space dust no matter how large the piece of space dust and there will only be darkness or there will only be emptiness or vacuum…

But that does not remain like that because now the dust which remained in the vacuum of space or in the vacuum of the universe must be taken out to give or to grant or to lend new knowledge to the vacuum of space or to the universe and that new knowledge will make a new beginning, in where there will not be any memory that there ever was a beginning before…

The manner in which existence takes the dust from the vacuum of space or from the universe is with black holes which truly are black vacuum cleaners.

The black vacuums or black holes also suck up any other matter or star or planet which has remained in the vacuum of space or in the space of the universe…

Once there no longer is matter or dust in space thus the black holes also will turn off or will stop from functioning because of lack of energy and they will disperse or they will disintegrate in the vacuum of space.

This emptiness in space, which now has become as if a vacuum or as if new space because of becoming as before the beginning thus will attract new knowledge in form of matter or stars or light or illumination…

And if in that new beginning the same happens which happened in the first, in where there was no renovation or there was no acknowledgement or no rebirth to continue on with life, thus also will have its end or ending even though it may take billions of years…

But that does not have to be as the above because as long as there are conscious beings in the universe, the universe has the very grandiose opportunity of renovating or of rebirth or of receiving acknowledgement so that the universe because of the conscious beings the universe will continue on without ever knowing end or ending...

Existence without knowing it renews every time she lends knowledge in the form of matter or in the form of stars or of light to the vacuum of space or to the universe and the universe cannot continue on for lack of acknowledgement or for lack of matter or for lack of renovated energy or illumination and thus the universe comes to its end or to its ending and thus making space for another beginning which will become as if the first beginning and also as if it never had an end or never an ending before...

But if the conscious beings are reborn or revive or take new life or receive acknowledgement through the very same knowledge or acknowledgement which they give or grant or lend, the universe will never ever have end or ending because the universe will continue on as if forever new and as if it never had any beginning...

Now then, every conscious being truly has the very grandiose opportunity of being reborn or of reviving or of taking new life or of receiving acknowledgement or rename to be able to continue eternally with life and in complete harmony and in complete abundance...

But if the conscious being does not desire that very grandiose opportunity of living eternally and living in complete harmony and in complete abundance with a higher or taller consciousness thus that conscious being only has to wait to die and that will he his end or his ending and nothing will become of that conscious being because eternity or immortality is not obligated or is not imposed to anyone or to anything, even a rock will stop from being or from existing…

Thus, one needs to be alive and conscious to be able to have or can receive immortality in the form of salvation and with her continue on alive renewing and renewing also everything else as savior…

Just as knowledge in the form of matter filled with energy or in the form of stars or in the form of light enters the vacuum of space or enters into the universe, thus that same way also thought or knowledge or illumination enters the conscious mind.

That thought or that knowledge or that illumination can take the conscious being to a great state or from one state to another state or to a greater identity or from one identity to another identity even though that conscious being really continues on with his physical form but every time that that conscious being enters into a greater state of knowledge or into a greater or new identity because of his knowledge, thus the physical form of that conscious being also is refreshed or is seen as if a new form…

But if the thought or if the knowledge or the new identity which enters in that conscious mind of that conscious being is a limited thought or is a limited knowledge or is a limited identity, thus that thought or knowledge or that identity, even though some type of energy or be it negative or be it positive, does not take that conscious being very far or into a greater state of identity and that thought or knowledge or that identity will disperse and the conscious mind becomes once again as if empty.

And if that conscious being nothing does with his conscious mind to have thought or knowledge or identity so that the thought or the knowledge or the identity takes him to a greater state or to a greater identity in where not only his conscious mind will be refreshed or becomes as if a new mind but also his physical form or body also will refresh or even could be cured from certain lacks or faults, such as of that of deafness in one ear or both and also some emotional lack or fault such as loneliness or shyness...

But if the conscious being in the course of his life does not enter into a greater state of thought or of knowledge or of identity, thus the conscious being keeps on dying until he completely dies and his body will discompose until it turns to dust and the conscious being has lost his very grandiose opportunity of rebirth and of continuing on with life as if new in complete or in perfect harmony and also in complete or in perfect abundance, perfect because it will be an abundance which will never ever end...

Now then, once the universe stops from discomposing or no longer the vacuum of space ever returns to nothing because of the conscious being coming to their maximum state or coming to their maximum identity and that way keeping the universe practically alive, thus there no longer will enter more knowledge in the form of matter or in the form of stars or of light because now that makes it possible the conscious beings because they will be the matter or the stars or the light or the illumination of the universe because of they being illuminated until the maximum or until perfection...

In other words, there will no longer be any more beginnings nor there will be anymore ends or anymore endings and the universe will become as if there were never beginning because of the universe becoming as if new or renewed for all of eternity...

And all the different parts of existence will act or will react as if one as the same the body and the conscious mind of the conscious being will act or will react as if one or as a single part and existence will be one with the conscious being because of the conscious being becoming or as being existence herself and reflecting through his body her glory…

Thus, when one as seed for more came out from the entrails of a man and one entered into the entrails of a woman, one had no memory of that exit or entrance even though one as a seed was in harmony and in abundance in those entrails and one also came out with all gladness and with all joy and also with all the good feeling of abundance and entered into the entrails of a woman and there also sought for knowledge of life to life receive and in her also enter and once in her also one forgot because once again one entered in harmony while one was transformed or one took the form of life which one did for or for the one which one received through the act of one or because of one's movement to find life…

And when one became complete in those new entrails, one humbled and one took the grandiose position of contender and one came out or one entered into the entrails of the world not only as much more but also one came out or one entered for much more and as much more.

But in the world one did not remember that one came out from harmony while one keeps completing the form of contender which could take one to not only come to be conscious but also which could take one to rebirth or a higher consciousness and continue on with life as if with a new form.

But if there were no rebirth which really is a higher consciousness because of lack of knowledge or because of lack of identity, thus that form not reborn would take one to death and that would be the end or the ending of all of her, life, and also of all of one…

Chapter Four

The More Understanding of Existence one has

The more understanding of existence or of reality one has of reality or of existence, therefore, the greater existence or reality will be to one as one will be to reality or to existence.

Reality or existence or the universe has everything to do with knowledge and acknowledgement, even unseen reality has everything to do with knowledge and acknowledgement even if negative knowledge or acknowledgement.

Space is an unseen reality or unseen knowledge or acknowledgement but space can really be acknowledged but only when there is knowledge to acknowledge space with.

In other words, space can really be acknowledged through or because of matter or light.

And matter as well as space can really be divided into three major parts.

Space is not only emptiness or is lack but space is also dark and cold.

These three parts which make space are lack or are negative (-) knowledge or negative (-) acknowledgement.

Positive knowledge or acknowledgement, therefore, is matter and its three major parts, such as weight or mass, light and heat.

And matter only interacts with space while in the vacuum or emptiness of space.

Now then, existence really is about knowledge and acknowledgement or confirmation and reconfirmation or on and off or stop and go or 01.

When matter which is knowledge enters the vacuum of space, space begins to react with matter as a form of acknowledgement and the vacuum of space begins to compress matter until matter explodes and she begins to expand in the vacuum of space.

But the friction which matters has from space makes matter to last less in the vacuum of space.

Now, before matter entered the vacuum of space, space was in a state of tranquility or space was neutral or 0.

But when matter entered the vacuum of space, space acted as if negative due to the interaction with matter such as the compression of matter but when matter exploded and began to expand through the vacuum of space, space began to act as if positive.

But when matter begins to turn off due to lack of energy, space also begins to act negative.

And when black holes begin to appear due to the collapse of large stars, the vacuum of space begins to act even more negative.

And when there no longer is matter in the vacuum of space because the black holes vacuumed all the matter up, the vacuum of space has become completely negative due to the black holes now expanding throughout the vacuum of space in search of matter.

But just as matter ran out of energy, so will the black holes run out of matter and collapse or disperse and the vacuum of space will once again become neutral or return to 0 and once again become ready to receive matter or new knowledge or illumination or 1.

But this new beginning is as if the very first beginning because there will not be any memory that there ever was any other beginning or beginnings.

Also, the vacuum of space is dimensional and will receive matter or knowledge or 1 up 118 times.

That is to say, there is 118 beginnings which really begun at the same time or spontaneously and enter into 118 dimensions or space time.

But not only that, matter or knowledge or 1 enters the first dimension or space time as 118 percent and that 118 percent equals or adds back to one.

In the same manner, 2 equals 236 percent and that 236 percent equals or adds back to one and so forth with the other numbers up to 118.

That is, 3 equals 354 percent and that 354 percent equals or adds back to one.

And finally, 118 or the last number equals 13924 percent, which is 118 times 118, and that 13924 percent equals or adds back to one.

The dimensions of time also add up as if they were one. And they increase in size from one to 118 times.

That is, the first is one but the second is twice as the first and the third is three times as the first but they still all add up as if they were just one dimension.

And if the speed of light was ever measured correctly the speed of light would be equal to one.

The speed of light is about 186, 282 miles per second which adds up to 9 or point nine or 09. This loss of 1 or of 01 is due to the compression of the vacuum of space.

It is similar to water when it freezes. When water freezes water loses energy and weights from 1 to point 9.

Thus, existence or reality is one or existence or reality is knowledge or more like 01 which really means knowledge or acknowledgement.

For the conscious being, therefore, this information really is very important because for this information the conscious being unknowingly struggles or unknowingly contends for to be able to enter into higher or taller or greater conscious modes or to be truly illuminated and through that illumination enter into a higher or taller or greater conscious existence and with this greater conscious existence or identity the conscious being can mode or transform existence or reality or his natural environment according to his will.

Of course, that forming or that transformation really is done through his mouth and through the use of the spoken word or the zero plus one factor or language, more like binary because of the presentation or the knowledge and the acknowledgement.

Chapter Five

The (0+-) Factor

Those that accept an idea or ideology blindly or without trying or testing that idea or ideology, thus they unknowingly become liars.

But nonetheless, liars they are and as liars they live and will lie to others to convince them to accept or believe and they will keep on lying until death herself and death herself as a black hole will close their lying trap.

Now, when one does the movement or gets interested for knowledge, thus one truly becomes that knowledge, true or not.

When the universe or creation came to exist, the universe or creation came to exist because of knowledge, but true knowledge even though most is unseen.

The proof is in the numbers or in matter which is composed of elements and the elements in turn are really numbers, real numbers.

In other words, matter is knowledge because matter is composed of elements and the elements are composed of numbers and the numbers are composed of positive, neutral and negative states, thus the, (+ 0 -), plus zero negative factor or a magnet with its three opposites parts.

What this really means is that knowledge can become neutral or void or useless and then become negative or contra productive if at first nothing was done with that knowledge, such as using or converting knowledge into acknowledgement or illumination or a positive or useful respond.

In the same manner as above, the universe or the vacuum of space becomes positive when matter or light enters the universe or the vacuum of space.

But as matter or light begins to burn out, the universe or the vacuum of space also begins to turn neutral until it turns negative, negative because of the black holes which now rule the universe or the vacuum of space which they now also suck up any remaining dust or matter or light to make space for another beginning.

But this new beginning is as if the very first beginning because there will not be any trace that there was ever a first beginning.

But the above matter or knowledge or illumination would only be a simple theory or an idea or ideology if the conscious being, which is also real knowledge, did for real acknowledgement or illumination and the conscious being would really have the power and the authority over matter or light to refresh matter or light and thus keep the universe or the vacuum of space always or for all eternity positive and refreshing in double abundance, all five portions of her!

Conclusion

Can you, the interested and the now the glad and the joyful reader, gladly and joyfully truly imagine what would had happened if the simple and the truthful or the real question as to the physical origin of the universe or as to the physical origin of all Existence, or even the truthful or the real question as to physical creation, were simply asked and then simply answered when conscious thought first entered the human conscious mind?

Really, the physical world or physical creation would not be the physical and the mental lack and the physical and the mental chaos that the world or that physical creation really is!

There would had never been insignificant or useless ideas or useless points of view that, unfortunately, still today bring forth false, half-truths or useless knowledge, not allowing mankind to further expand his conscious mind into simple enlightenment or simply into completeness or perfection, and therefore, not allowing the future or complete or perfect Existence or perfect creation here and now.

Unlike the divisible atom, where the electrons cannot naturally return to the center or to the nuclei of the divisible atom or where the proton naturally cannot occupy electron rings or the electron field or shield, and certainly the physical universes would really be physically different and therefore all of actual Existence, the conscious mind or the conscious living being can deter, can reverse, can change or really can improve the conscious mind's or the conscious living being's point of view, believe or understanding or the way of conscious thinking or the way of conscious functioning or even the way the conscious living being learns simply by changing or improving conscious thought or even by improving mood or by simply acknowledging or by simply seeking to be enlightened or illuminated or to be reborn, which really is a higher or taller consciousness in harmony!

By getting really interested or by improving one's mood is a very simple way to improve one's conscious thought or one's point of view.

One can see more by simply getting interested or simply by being glad or joyful or even excited!

Getting interested or getting glad or joyful or even getting excited is a major part of learning, but getting interested or getting glad or joyful or even getting excited is not an automatic process.

The conscious living being also has to make an effort to get interested or to get glad and joyful or even excited.

In other words, the conscious living being really has to make an unnatural conscious effort to really learn how to really learn.

Once the conscious living being simply or actually knows or simply or really understands how to learn or really understands the real reason for really learning, which is really recollecting or simply remembering, thus the conscious living being will know or will understand how.

That is to simply say, once the conscious living being really knows or really understands why or really understands what or really understands the simple reason for getting interested or the simple reason for getting glad or even excited, the unnatural process of getting interested or the unnatural process of getting glad or even getting exited then becomes automatic.

And the conscious living being through conscious thought or through conscious effort will really or will actually know how (to complete Existence or how to complete actual reality or physical creation).

Interestingly, joy, gladness or happiness, as really is also conscious thought, is pure energy.

A joyful, a glad or a happy thought can really or can actually keep the conscious living being a good number of days without sleep, without hunger and without feeling tired or stressed out.

Only when the joyful, the happy or the interested thought is gone or completed or has actually become natural, the conscious living being once again begins to need sleep, to need food and to need rest, both physically and mentally.

True, complete or real joy; true, complete or real happiness; or true, complete or real interest really brings to light physical things or the reality unseen.

True, complete or real joy or real happiness is not an automatic or is not a natural process at first, even though true, complete or real joy already exists.

And because mankind instinctually or naturally knows that true joy or that true happiness is or already exists, mankind seeks true joy or true or real happiness.

However, that is temporary or that is natural joy or that is temporary or natural happiness.

That is the reason that mankind really feels miserable or really gets into trouble looking for temporary or instant or natural joy or temporary or natural happiness.

True, complete or real joy or happiness is unnatural, and therefore, really requires conscious effort or conscious thought.

However, this really has nothing to do with positive thinking.

Positive thinking is only a very simple point of view. Positive thinking is not really or actually a good thing.

Positive thinking can really be a real hell or a very heavy burden. Mankind usually, as a creature of unnatural habit, imitates mostly what mankind really likes or what really brings mankind the most pleasure or the most joy.

Changing or improving conscious thought, changing or improving behavior or changing one's way of thinking or changing point of view is unnatural and is also not very easy.

It really requires a lot of effort, a lot of energy and a lot of time for a lot of nothing.

Positive thought really is, therefore, one of those good sounding lies or good sounding half-truths or false science.

Positive thinking bears all as it pretends to dress and nurture.

But once the conscious living being or mankind through conscious effort really acknowledges, realizes or really understands or really recollects the real or the actual reason for true, complete or real joy or real and lasting happiness, then true, complete or real joy or real and lasting happiness simply becomes an unnatural but automatic process as conscious thought once became an unnatural but automatic process as the very simple magnetic field or the very simple neutral shield always was an automatic process of physical existence or of physical reality.

In short, once the unnatural but automatic process of true, complete or real joy or real and lasting happiness kicks in, new energy is felt both in body and in mind.

At first that new energy will feel like an ecstasy or a natural super high.

In fact, it is really or actually the first enlightenment or illumination from above!

It is a true, a complete or a real joy or a real and lasting happiness that keeps coming, even without further effort from the conscious living being, leading to a second and final enlightenment or final acknowledgement.

The very simple anticipation of the next enlightenment really or actually is what makes possible the new found joy or real and lasting happiness.

Ironically, when true, complete or real joy or real and lasting happiness begins automatically as once did conscious thought, the conscious living being really does not understand where it really comes from and so the conscious living being begins to fight real and lasting happiness, as the conscious living being once fought against conscious thought, by feeling odd or by feeling strange, and thus begins to reject that new and automatic and real joy or real and lasting happiness.

And in conclusion or in continuation, the most important thing in all of creation or in all of Existence, seeing or not seeing, once again, simply is or actually is but conscious thought!

Conscious thought only exist in conscious living beings. Conscious thought really is not a natural or is not an automatic process as is all of Existence or as is reality.

Even all of Existence or all of creation as well as non-existence really require something.

Conscious thought is not a natural process and conscious thought really requires conscious effort from the conscious living being.

Even though knowledge simply exists or simply "the fact simply and already but is," the conscious living being does not know or does not learn or does not really recollect or does not really remember and thus simply cannot complete Existence or physical creation until the conscious living being makes a conscious effort to simply think and to simply ask that unnatural but true or real question which will simply bring the conscious living being complete or perfect knowledge or acknowledgement or enlightenment!

And the real or the actual purpose of complete or perfect knowledge is simply or really to let the conscious living being know or really acknowledge that the conscious living being but is and that the conscious living being can really do more with what the conscious living being could really be or could really recognize or could really recollect or could really remember!

Existence as knowledge simply is or simply exists. But ironically, Existence or knowledge knows not.

However, just because all of Existence or all of knowledge really knows but not, really does not make Existence or knowledge dumb or useless.

All of Existence or all of knowledge acts or reacts through a simple magnetic or through an electronic impulse or acts or reacts through an actual magnetic instinct.

But knowledge does not know that knowledge really is or that knowledge simply exists.

For example, the simple apple tree grows knowing not; but the tree's natural instincts, through knowledge already gotten or already present, is to produce a seed or to simply multiply itself.

The apple of that apple tree ripens or temporarily becomes almost complete not knowing that it does, so that the seed in the almost complete apple can be carried away, but not by the wind or not really by rolling on the ground even though the apple is almost round, but really carried away by living beings.

That is the very simple or the actual reason for the sweetness or the sugar in the almost round apple!

The simple apple in the belly of the living beast or in the hand of the living being is simply or is really a continuation, and not the conclusion, of something that already was.

And even though the apple tree which is really but simple knowledge was really before the seed, the purpose of the apple tree was always the seed, an extension or an expansion of physical knowledge that really already was or was already present!

Another very interesting thing about the knowledgeable apple seed is that the apple seed not only carries all of the physical information of the apple tree, but also the apple seed carries all the knowledge or all the information about the environment or the physical surroundings where the apple tree was or is.

The knowledgeable apple seed no longer has to experiment with the environment or with natural surroundings since the knowledgeable seed already has that physical information.

Instead, the seed as a new or as a second tree will gather more or new knowledge for the next apple or the third tree not yet present.

Whether the new or the second apple from the new or the second tree is sweeter or is more sour than the very first or the previous apple, simply depends, of course, on the physical reaction or the interaction that the living beast or the living being gave to the very first or to the previous apple.

The knowledgeable apple is often colored red to attract the eye of the living beast or the living being with hands, and the redder or the more reflective or shiner the apple, the better for the apple tree and thus the better for the knowledgeable apple seed.

If the natural or the physical surroundings are all red, then the knowledgeable apple with time simply changes to another but more attractive or more attracting color.

By the way, animal seeds or animal cells as do plants seeds have the very same ability or instincts to carry knowledge of previous generations.

But because of the animal's irrational or emotional behavior, or even point of view or new physical location, that first or that previous knowledge or instincts is rarely re-used and usually is forgotten, rejected or even lost.

A very good or an excellent example of lost knowledge or lost instinct is the domestic animal. Left alone in an unfamiliar environment the domestic animal will not survive.

Even house plants lose the ability to survive because of the lack of information or the lack of knowledge of the new environment or new surroundings.

Once again, conscious living beings are really the most important thing in all of Existence, and that is what Existence really reflects or that is what all of Existence really wants or really requires the conscious living being to know. But not only that!

All of Existence also requires, also wants, or also needs, as the apple tree, the conscious living being to simply act or to react to knowledge, so that all of Existence as well as the living conscious being can be further completed or perfected through conscious thought.

In a way, conscious living beings are actually or are really inside a very large mechanical or infinite magnetic brain or inside a very large electronic brain; and conscious living beings are really the thinking brain cells or living knowledgeable seeds or even living knowledge processors.

Ironically, Existence herself will not be complete or will not be perfect until all the needs of the conscious living being are simply but met through conscious thought.

And that is the actual reality!

The only true or the only complete or the only perfect or real beginning in Time and in all of Existence or even creation herself is the simple or the actual beginning of conscious thought.

The only limits that the human race, the conscious living being or that mankind really has are the very same limits that the human race, the conscious living being or that mankind really gives or really sets upon him very self.

So, therefore, let's get glad and joyfully and continue on to think and rethink!

Thinking or conscious thought will one day keep the universes or creation as well as mankind from collapsing or dying.

Conscious thought will really be the conscious shield that will keep the simple universes or creation her very self from collapsing or from imploding into a giant black hole or into a giant grave or into the final end.

Conscious thought will be the conscious shield or the conscious energy that will one day really keep the simple human conscious mind or the human brain from physically collapsing or from physically imploding also into a hole or into a grave, thus saving the physical universes or saving creation and thus completing or really perfecting physical Existence or physical creation!

The final reality of the matter is that as long as there is conscious thought in the physical universe or physical universes, the physical universes or physical creation will not collapse or will not implode or will not even turn on itself.

The reality of the matter is that there was only one physical creation or only one physical beginning, actually a physical continuation of what already was.

As long as there is conscious thought, only one physical beginning or only one physical continuation or only one physical creation will be so until infinity!

Further Notes

Just as the black hole removes all dead matter from the vacuum of space or from the universe to make the vacuum of space or the universe neutral to attract need matter or light or illumination from outside the vacuum of space or outside the universe, thus we should do the same with our conscious mind with our dead beliefs or with our dead faith which will take us dead to the grave if we do not remove our dead beliefs or our dead faith and that includes blind faith from our conscious mind...

Seek the Truth

Seek of the truth or seek the truth and when she let herself be found from you, you truly will become taller, bright and will even have more confidence in yourself.

The very curious thing about the truth is that she begins with one and if one truly does the movement to find her, thus she will come to one or she will be given to you by he that made or that presented her for the truth.

A Higher Mental Consciousness

The path of less resistance or without contention not only leads to laziness or to vanity or to nothing but also the path of less resistance or without contention leads to death.

And death ends any possibility to any other possibilities, such as a higher mental consciousness or illumination with real gladness and with real joy, in double abundance all five portions, and also with the power and with the authority of the heavens here on the earth.

Who am I really?

My pen name as a writer is Forester de Santos and I am truly on a very grandiose crusade of rebirth alive or to be born again into royal life, which really is a higher mental consciousness, with complete gladness and with complete joy and even also with complete abundance of God but as much more than God and as much more than Creator. God equals true or real knowledge or true or real acknowledgement, which can only be achieved through a higher consciousness...

Now then, one who truly is on a very grandiose crusade cannot follow another or he cannot let himself be surrounded by his beloved ones or his fans because he cannot cross over them or he cannot cross over because of them being in the way or because of them blocking the path to righteousness which really is but which cannot be seen until rebirth or until one is born again or until one enters a higher or taller mental consciousness, which really is an expansion of mental consciousness or self-awareness with complete gladness and with complete joy and even with complete abundance.

I do not, therefore, ask to be followed, not because I will not lead, but because I will not look back, because I am not going that way, but as I move forward and ahead I will only look to my right and to my left to see who walks with me.

But so far only one walks besides me, at times in front of me and often on my left side…

But those that truly decide to follow me will become as me and as me will truly receive or truly gather true knowledge because my struggle or my contention or my very grandiose crusade of rebirth is true, so true in fact that I have become a much better person because of the true faith which I have come to receive through my search and research for the truth.

And because I have come to have true faith or faith of God, which is true knowledge, thus I use my true faith as a shield to repel or to reject other beliefs or good sounding lies that will only take one dead to the grave!

Therefore, to rebirth alive or to be born again, really a higher mental consciousness, while still living here on the very earth which will be as in the very heavens through rebirth!

Now then, rebirth or life renewed is really about a higher mental consciousness here on earth!

$$(- (- (- (- (- (- (- (0 + 1) +) +) +) +) +) +)$$

If you truly enjoyed this simple and humble work, please leave a comment according to your good pleasure and give also a rating but also according to your good pleasure.

Thanks so very much for your time and best of wishes, Forester de Santos.

Thanks for reading my work!

0+1 = peace and knowledge to all mankind...